A Simple Solution

To a

Complex Problem

By

Dow Edward Hendricks

To Meak,
I hope you enjoy it.
  Ed

Copyright © 2020 by Dow Edward Hendricks

Dow Edward Hendrick, Publisher, Parma, Ohio 44134

ISBN-978-1-79489-316-0

All rights reserved. Printed in the United States of America. This publication is protected by Copyright and permission should be obtained in writing from the publisher prior to any prohibited reproduction, storage in a retrieval system, or transmission in any form or by any means, electronic, mechanical, photocopying, recording, or likewise.

First Edition February 2020

Hendricks, Dow Edward

A Simple Solution to a Complex Problem/Dow Edward Hendricks, 1st Edition 2020

Introduction

Climate change.

There are few topics that generate more passion than climate change. For the average person, the truth about climate change is shrouded in pseudo-science, political agendas, and misinformation. The truth is whatever your favorite politician, or political pundit wants it to be. Politicians will spin same set of facts, or the same scientific study to suit their party's position.

Let me get this out of the way right now. Politicians lie about climate change and global warming. It is nearly impossible to really understand what is happening to our climate, unless you, as an individual, do the research yourself.

For example, President Obama talked about the University of Illinois' study that said 97% of scientists believe in global warming, and that 82% said that human activity was the main contributing factor (1). On the surface that sounds horrible for our world. It must be true since the President of The United States referenced the study. Right?

Wrong! President Obama outright lied to the American people about this study. He twisted the results of the study to meet his political agenda. Let me explain the study's results, without a political agenda.

In 2009, Peter Doran, University of Illinois at Chicago associate professor of earth and environmental sciences, along with former graduate student Maggie Kendall Zimmerman, conducted the survey(2). They sent out a two-minute online survey to 10,257 earth scientists. The survey consisted of two questions. The first: "When compared with pre-1800 levels, do you think the mean global temperatures have generally risen, fallen, or remained relatively constant? The second: Do you think human activity is a significant contributing factor in changing mean global temperatures?"

Of the 10,257 surveys sent out, 3,146 people responded, or approximately 31%. Of the 3,146 respondents, only 79 people were actual climate scientists. In reality, 97% is a false number. Actual climate scientists made up .77% of those who responded to the survey. In other words, less than 1% of those to whom the survey was sent stated they felt global temperatures are rising and that human activity is a significant contributing factor.

Climate change and global warming have become big business. The Paris Climate Accord; Al Gore's an Inconvenient Truth; The Heartland Institute; President Obama; The Green New Deal; The United Nations; and many Congressional Representatives have made climate change a major issue.

Melting Ice Caps

Whether or not you believe climate change is real and that humans significantly contribute to climate change one fact remains: from the 1970s to today NASA data shows the polar ice caps are melting. Although, not in a uniform manner. The Arctic ice cap seems to be more distressed than the Antarctica ice cap (3).

Data seems to indicate there are two factors responsible for the melting ice caps. The first factor is the earth is coming to the end of an ice age. Global temperatures are rising as part of the natural climate process. The second factor is that humans are in part, but not entirely, responsible for the increase in greenhouse gases and rising global temperatures.

Another irrefutable fact is when the ice caps have completely melted, a lot of water will be returned to the oceans. How much water? To obtain that number some calculations are in order.

Today, the earth has approximately 9 million cubic miles of ice (4). One cubic mile of ice contains slightly over 1.1 trillion gallons of water. If all that ice melted, around 10 quadrillion gallons of water would be returned to the oceans (5). That's a big number. How much is 10 quadrillion gallons?

It's difficult to visualize 10 quadrillion gallons. For example, The Great Lakes hold 6 quadrillion gallons of water (6). However, we need a better example to help visualize 10 quadrillion gallons of water.

Let's say we could build a sturdy wall around the United States. Let's make the wall 17 feet high just to be safe. Then, let's drain the entire Great Lakes and let that water flood the continental United States. The entire United States would be flooded to a depth of over 9 1/2 feet. If that amount were 10 quadrillion gallons, instead of 6 quadrillion, the entire United States would be flooded to a depth of 15.2 feet.

The oceans are going to get a lot more water added to them, which will cause them to rise. The dispute among scientists is how much will the oceans rise. By the year 2100, conservative estimates have the oceans rising by three to five feet. Less conservative estimates have the oceans rising by 16 to 20 feet.

If ocean levels rise this much the world geography is going to drastically change. Let's look at how rising ocean levels are going to effect the United States. The following states have their lowest point at or below sea level (7).

>Delaware
>Florida
>Louisiana
>Rhode Island
>New Jersey
>Mississippi
>South Carolina
>Connecticut
>Alabama
>Massachusetts
>Georgia
>Maine
>North Carolina

New York
New Jersey
Pennsylvania
Texas
Washington
Oregon
Alaska
California
Virginia

Many nations and cities throughout the world will be effected by rising ocean levels. To name a few: the Bahamas; Bangladesh; Osaka, Japan; Rio De Janiero; Miami; The Hague; Hong Kong; and Shanghai might cease to exist. Much of New York City and Long Island would be under water.

The Secret That Politicians Do Not Want Us to Know

Irrespective of the cause, ocean levels are rising. What our political leaders do not want us to know is that we cannot stop climate change. We cannot stop the oceans from rising. Not one single tear from Congresswoman Alexandria Ocasio-Cortez, or from former Vice-President Al Gore will stop the melting ice caps. Adjusting human activity can only slow the rate at which the oceans are going to rise, and not by much.

The Green New Deal; The Paris Climate Accord; limiting carbon emissions; eating less meat, preventing cows from farting; New York City School kids having meatless Mondays; or any other current proposals will not stop the oceans from rising and causing mass destruction. Let me repeat this because it is important: **there is nothing human activity can do to stop the oceans from rising.**

One might ask, if there is nothing, we can do to stop the oceans from rising, why are politicians acting out? What a wonderful question. In theory, politicians primary concern should be keeping their citizens safe. Politicians are not supposed to appear totally inept and helpless. Even when they cannot solve a problem, they still must act like the problem has a solution.

In the United States, we have politicians who enjoy feeling that our country is the cause of all global pollution, greenhouse gas emissions, and melting ice caps. President Obama felt that way, as does Congresswoman AOC, Senator Sanders, Senator Warren, former Vice-President Al Gore, most of the United Nations, and countless others. Yet, none of these individuals and entities have proposed a workable, affordable solution to rising ocean levels and greenhouse gas emissions.

In reality, the United States is not the world's largest polluter or environmental offender.

A Simple Solution to this Complex Problem

The phrase *simple solution* is misleading. The scientific solution, the physical solution to the rising oceans is already available. We have the technology to prevent this disaster from taking place. Also, when the solution is implemented the problem with greenhouse gases will be solved.

I want to emphasize this statement. **The solution to rising oceans and excess greenhouse gases already exists**. The solution lies in existing technology, not in legislation. The solution does not come from eating less meat; stopping cows from farting; electric cars; ending fracking; stopping the use of coal; stopping dependence on oil; or any of the many other suggestions.

Although we have the technological means to solve this crisis, the nations on earth need the political will to implement this solution. And, that is a problem. For decades, especially in the United States, we have heard groups like Green Peace, The Sierra Club, and countless other environmental organizations tell politicians something must been done to combat global warming. In the process, global warming has become big business.

In the United States, depending on your political party, global warming is going to kill all life on earth, or it is the biggest hoax every perpetrated on our citizens. Politicians use the same studies and interpret those studies differently in order to achieve their political goals. This needs to stop if a real solution is to become effective.

Every country on earth will need to participate to end this crisis. Most countries signed the Paris Climate Accord, with only Syria and Nicaragua abstaining, and now the United States (8). Despite protests from environmentalists, the United States is not the world's biggest polluter. Depending on the study, China and India do more to pollute the atmosphere and oceans than The United States. Russia and Japan round out the top five worst polluters (9).

The solution to rising oceans and greenhouse gases is sure to irritate many environmental groups. These groups protest very well, but they do not offer a real solution, other than the United States is evil and we must stop driving cars, eating meat, drilling for oil, and fracking. In truth, the United States has done more, without The Paris Climate Accord than China, India and Russia combined.

The solution is available. However, it is worth mentioning again that every country on earth needs to participate. The United Nations must support the solution. Finally, environmental groups are certain to protest the solution for multiple environmental impact reasons. However, they really need to realize this simple solution will save millions of lives, create millions of jobs, mitigate problems caused by rising ocean levels, remove greenhouse gases from our atmosphere, provide fresh drinking water to every person on earth, and end droughts forever.

In essence, this simple solution will save our planet.

Does this sound like fantasy? Perhaps. Especially, if scientists, politicians, environmentalists, and the big business climate change people want to continue with climate change fear mongering. These people and groups need to understand that the solution to rising ocean levels, greenhouse gases, and climate change is within our grasp. The solution can be implemented quickly, and the effect will be immediate.

The Solution

The inspiration for the solution to rising ocean levels and greenhouse gases came from researching man-made lakes. Man-made lakes can be found all over the planet. Essentially, anywhere a dam is built, a man-made lake can be created.

There are two ways to measure the size of a man-made lake. The first way to measure a man-made lake is by surface area. Lake Volta in Ghana covers 3,725 square miles, which makes it the largest man-made lake by surface area (10).

The second way to measure the size of a man-made lake is by the volume of water in the lake. Using this method, Lake Kariba is the largest man-made lake and it contains over 12.5 trillion gallons of water (11). Lake Volta holds 10.3 trillion gallons of water. By comparison, Lake Mead holds 9.45 trillion gallons of water.

The Great Lakes cover an area of 94,250 square miles. This is about the size of Wyoming. The Great Lakes hold approximately 6 quadrillion gallons of water, or 60% of what is going to be released by melting ice caps.

While researching man-made lakes I began to wonder if it were possible to build man-made lakes that could hold the 10 quadrillion gallons of water that will be released from the melting ice caps. Could mankind prevent the rising ocean levels from unleashing unimaginable destruction?

The answer, surprisingly, was yes. We would need over 150,800 square miles to make this work. Fortunately, there are several nations on earth that have the land area to make the made-made lakes which could hold 10 quadrillion gallons of water. Excess water from the melting ice caps could be piped into these man-made lakes.

The Sahara Desert is 3.5 million square miles (12). By comparison, the United States is 3.7 million square miles. In other words, the Sahara Desert is approximately the size of the United States.

The Arabian Desert is 900,000 square miles (13). The Gobi Desert is 500,000 square miles (14). The Kalahari Desert is 360,000 square miles (15). The Great Victoria Desert is 220,000 square miles (14). Finally, the Pantogonian Desert is 200,000 square miles (16). The total area of these deserts is 5.680 million square miles: far more than the 150,800 square miles needed.

The Sahara Desert lies in parts of seven countries: Morocco, Mali, Mauritania, Egypt, Libya, Algeria, Chad, the Niger Republic, Nigeria, Sudan, and Burkina Faso. It runs 3,000 miles in an east to west direction and 800 – 1,200 miles in a north to south direction.

The Arabian Desert covers most of Saudi Arabia. However, it also extends into Jordan, Iraq, Kuwait, Oman, UAE, Yemen, and Qatar. It is approximately the size of France.

The Gobi Desert is in China. It measures 1,000 miles east to west and 500 miles north to south.

The Kalahari Desert 1,000 miles north to south and 600 miles east to west. It occupies nearly all of Botswana, a third of Eastern Namibia, and the northernmost region of North Cape Province, South Africa.

The Great Victoria Desert lies in Australia. It measures 434 miles east to west and 372 miles north to south.

The Pantogonian Desert lives primarily in Argentina. It is 1,968 to 2,624 feet above sea level. It measures 900 miles north to south and 150 miles east to west.

These six dry desert locations have the potential for the world to create the man-made lakes, which could hold the 10 quadrillion gallons of water which the melting ice caps are going to release. Putting aside, for the moment, the physical and construction challenges of this task, there are political and religious concerns to overcome.

There would be fewer political issues to overcome with building man-made lakes in the Gobi, the Great Victorian, and the Patagonian than building in the other locations. Each of these deserts is located in one country, with one political system. There would be fewer religious difficulties to overcome as well.

China is a one-party government. In theory, the Chinese constitution guarantees freedom of religion. If the Chinese government thinks the religion is in anyway anti-government, then the Communists brutally ban the religion. The religious affiliation in China breaks down as follows (17):

    Atheist - 61%

    Tao or Confusion Philosophies - 26%

    Buddhist - 6%

    Christianity - 2%

    Folk Salvationist - 2%

    Islam - 2%.

Religions in Australia are as follows (18):

    Roman Catholic Christian - 25.3%

    Atheist or Agnostic - 22.3%

    Christian other than Catholic or Anglican - 18.7%

    Anglican Christian - 17.1%

    Other Beliefs - 10.1%

    Buddhism - 2.5%

    Islam - 2.2%

    Hinduism - 1.3%

    Jewish - .5%.

Finally, religions in Argentina are as follows (19):

    Catholic - 65%

    Atheist or Agnostic - 20.6%

    Protestant - 11.7%

    Other - 1.7%

    Other Christian - .7%

    Judaism - .2%.

  Religious issues become a problem in the countries where the primary religion is Islam. The two dominant Islamic sects are Sunni and Shia. There is a history of difficulties between these two sects that dates back to the death of Mohammed.

  Saudi Arabia's population is 75% Sunni (20). Iran, by comparison, is 90% Shia (21).

The Arabian Desert lies primarily in Saudi Arabia. Saudi Arabia is 756,984 square miles. One-third, or 249,804 square miles is desert. Saudi Arabia has five major oil fields, which lie within an area of approximately 5,200 square miles. This is an area slightly larger than Connecticut.

The point being made, is there is plenty of room in the Saudi Arabian portion of the Arabian Desert to build a man-made lake. Again, for comparison, The Great Lakes have a total surface area of 94,250 square miles; are 5,472 cubic miles; and hold 6 quadrillion gallons of water.

Thirty-seven percent of the Arabian Desert would be needed to build a man-made lake the size of the Great Lakes. This single man-made lake could hold 60% of the water that is going to be released from melting ice caps.

Saudi Arabia would then have one of the world's largest fresh water sources and one of the world's largest reserves of oil. With an abundant source of fresh water, this desert area could become fertile. Trees could be grown, and forests created, which would remove greenhouses gases from the atmosphere. Dams, like the Hoover Dam could provide electricity to a vast area of the middle east. All this from one man-made lake.

Similar man-made lakes could be built in the other large desert areas throughout the world. The Sahara Desert would be a logical place to place to build a man-made lake. To build a man-made lake, or several man-made lakes in the Sahara Desert is possible. However, religious, political, and environmental difficulties must be dealt with.

To bypass many of the political difficulties associated with building a man-made lake in the Sahara Desert, such a lake could be built solely in Egypt. Egypt is, in theory, a democracy, and the government is somewhat stable. Egypt's primary religion is Sunni Muslim (20). Christians make up the other 10% (22).

The Sahara Desert covers 300,000 square miles of Egypt. One-third of this area could be devoted to a man-made lake. With this abundance of water, forests could be grown over the other 200,000 square miles of Egyptian Desert.

The Gobi Desert covers 500,000 square miles, or 320 million acres. Another man-made lake could be built here.

Forests

The earth has approximately 56,650,000 square miles of land (23). Approximately, 30%, or 16,995,180 square miles of the earth's land mass is covered by forests (24). This means there are 10.877 billion acres of forest on the earth.

Humans put 40 billion tons of CO2 into the atmosphere every year (25). One acre of forest can absorb 2.5 tons of CO2 annually (26). The forests on earth absorb around 30% of human emissions, or 12 billion tons, which leaves 28 billion tons of CO2 in the atmosphere. Is it feasible for new forests to absorb 28 billion tons of CO2?

Let's go back to Saudi Arabia. If a man-made lake were built in Saudi Arabia covered 94,250 square miles, 150,354 square miles of desert would be left. Let's say 100,000 square miles, or 64 million acres, were used to grow new forests. This new forest would eliminate 160 million tons of CO2 from the atmosphere.

The Egyptian portion of the Sahara Desert, after their man-made lake was built, would have 200,000 square miles, or 128 million acres of desert left. These 128 million acres of new forest would remove another 320 million tons of CO2 from the atmosphere.

A new, 200-million-acre forest in the Gobi Desert could absorb 500 million tons of CO2.

Just three new forests will absorb 980 million tons of CO2. There are several other areas where new forests could be grown. It is possible for new forests to absorb approximately 2 billion tons of CO2. While it helps, it is far short of the 21.754 billion acres that are needed.

The United States has an area of 2.3 billion acres. Only 66 million acres are considered developed. That leaves 2.234 billion acres of undeveloped land. Forests planted on 500,000 million acres would absorb 1.250 billion tons of CO2.

China is about the same size as the United States. So, new forests planted in China could absorb another 1.2 billion tons of CO2.

The planting of three new forests in deserts and new forests in additional two countries would absorb 3.48 billion tons of CO2 annually. While this is a step in the right direction, we still need to rid the earth's atmosphere of 18.274 billion tons of CO2.

We have not looked at the possible size of a new forest in every country. India, Russia, and Australia have vast areas where new forests could be grown. Also, every country on earth has areas where new trees could be planted. If every country participated in planting new forests, it is possible for all excess CO2 to be absorbed by trees.

Electric Cars

As Democratic candidates, campaign for their party's 2020 Presidential nomination, several candidates have clearly stated they want to destroy the fossil fuel industries in America. With no consideration what that destruction would do to the American economy they want to stop fracking, coal mining, and oil production. This would raise the price of gas, force people to drive private vehicles less, and increase the use of electric cars.

What these presidential candidates fail to acknowledge are market and technological forces. Market forces and advances in technology will make gas powered motor vehicles obsolete within the next two decades.

Electric cars are the future in driving. According to Stephen Leahy in his September 13,2017 article in National Geographic Magazine, by 2040, 90% of all motor vehicles in the U.S., Canada, Europe will be electric. This will cut CO2 emissions by 3.2 billion tons a year.

China, the largest contributor of atmospheric CO_2, has 20 different electric vehicles currently available. According to Wang Chuanfu, chairman of Chinese automaker BYD Co., Ltd. by 2030 100% of vehicles sold in China will be either full-electric or mild hybrid vehicles.

India wants 100% of cars to be electric by 2030. The goal, while laudable is unrealistic. Most experts have a more realistic goal of 40 – 45 % of electric cars in India by 2030.

The current estimated population of the United States, Canada, Mexico, and Europe is slightly over 1.2 billion. The current estimated population of China and India is 2.7 billion, or more than double the combined population United States, Mexico, Canada, and Europe.

If China and India reach their goals of 100% electric cars by 2030, CO_2 emissions will be further reduced by approximately 6.4 billion tons per year. So, by 2040 with the dominance of electric cars, CO_2 emissions will be reduced by 8.6 billion tons per year.

The point is that electric cars will make the fossil fuel industry nearly irrelevant by 2040. The panic by American politicians and politicians worldwide is unwarranted. Market forces will help solve the greenhouse gas problem.

The Cost

At the time of their construction, in today's dollars, the estimated cost of building Lake Mead and the Hoover Dam is between $700 - $942 million (27). Compare this amount to the Department of Transportation recently awarding San Diego $2.1 billion to extend their trolley system by 10 miles (28).

Much like the facts of global warming, the costs of building the solution can, and will be manipulated. Cost over-runs, without an independent watchdog will soar. But, the cost of doing nothing will be astronomical. Also, the cost of the solution will be far less than the costs of implementing Congresswoman AOC's Green New Deal, which is estimated to be up to $93 trillion.

At the time of its construction in 1965, Lake Volta cost $258 million. In today's dollars, the cost of building Lake Volta would be slightly over $2 billion. Lake Volta hold around 39 trillion gallons of water.

At the time its construction was completed in 1977, Lake Kariba cost $480 million. In today's dollars that would be around $2 billion. Lake Kariba holds around 47 trillion gallons of water.

Together these two man-made lakes hold 86 trillion gallons of water. Their construction costs in today's dollars is $4 billion. To save our planet, we need man-made lakes that will hold 116 times the amount of water held in these two lakes.

The math is fairly straightforward. The cost to build man-made lakes large enough to hold 10 quadrillion gallons of water, would be approximately $464 billion, in today's dollars.

Along with man-made lakes, desalination plants would be built to provide fresh drinking water. In 2014, Ras Al-Khair in Saudi Arabia went online. Construction costs, in US dollars were around $6 billion (29). The plant produces 160 million gallons per day of fresh drinking water.

For comparison, a recent desalination plant in California expected construction costs were $329 million. The plant's construction and operating costs for a 30-year period were expected to top $1 billion. California has received pushback because of the costs and the second desalination plant's future is in doubt (30).

Logically, California makes no sense. Over the course of 30-years, it would cost approximately $33 million per year to operate a desalination plant. This works out to less than $1 per person per year in California. For comparison, a July16, 2014 news story by NBC, stated the drought in California cost farmers $2.2 billion in losses, and resulted in 17,000 lost jobs. In 2015, droughts cost California farmers $1.8 billion in losses, and resulted in a loss of 10,000 jobs. In 2016, drought cost California farmers over $600 million in losses, and a loss of nearly 2,000 jobs. In just three years, drought cost California farmers over $4.4 billion in agricultural losses, and 29,000 lost jobs (31).

If operating full-time, desalination plants in California would end droughts in the state. Also, fresh water could then be supplied to Lake Mead, which is at historic low levels. California could be a major resource of fresh water to the entire drought stricken southwest, if their politicians had the political courage to take sensible action.

The cost of planting forests is much less. Estimates vary, but the average cost of planting one acre of seedlings is $400 - $500 (32). We estimated 1.3 billion acres of new forests could be planted around the new man-made lakes. The estimated cost of planting these forests $696 billion.

This sounds like a lot of money: and it is. For comparison, let's look at some military budgets. In 2019, the following countries are the top ten in military spending (33).

#1	United States	$717 billion
#2	China	$177 billion
#3	India	$61 billion
#4	Germany	$53 billion
#5	Saudi Arabia	$51 billion
#6	The UK	$49 billion
#7	France	$48 billion
#8	Japan	$47 billion
#9	Russia	$46 billion
#10	South Korea	$42 billion.

These ten countries spend a combined $1.2 trillion per year on the military. The total cost of building this simple solution for saving our planet will cost between $1 - $2 trillion.

This is worth repeating: in total, to build man-made lakes, plant new forests, and build desalination plants will cost between $1 - $2 trillion. This works out to a lifetime construction cost, for every person on earth of $250 per year (annual operating costs have not been calculated). This number is small compared to the estimated cost of the devastation rising ocean levels are going to cause.

Conclusion

The estimated cost to resolve the problems caused by rising sea levels and excess CO2 in our atmosphere will be between $1 - $2 trillion dollars. Compare that cost to the estimated cost of The Green New Deal, which is between $93 - $100 trillion.

Rising sea levels are expected to cost the world economy $21 trillion per year by the year 2100.

If we, as the people who inhabit the planet earth act now, we can save our planet. We can save the planet at a fraction of the cost of doing nothing, or of trying to implement some insane, unworkable scheme, like The Green New Deal. We can add millions of jobs, save millions of lives, provide clean drinking water to millions, and create beautiful, new growth forests and lakes.

To implement this solution, technical and potential environmental issues will need to be resolved. Issues like, what types of materials will be used in construction; and the environmental impact on wildlife. Those issues can be overcome.

We simply need the political strength and will on a global scale to achieve this goal. Failure to do so will be catastrophic.

References
1. https://www.forbes.com/sites/alexepstein/2015/01/06/97-of-climate-scientists-agree-is-100-wrong/#4a2517143f9f
2. https://www.eurekalert.org/pub_releases/2009-01/uoia-ssa011609.php
3. https://www.nasa.gov/content/goddard/nasa-study-shows-global-sea-ice-diminishing-despite-antarctic-gains
4. https://www.answers.com/Q/There_are_about_9_million_cubic_miles_of_ice_on_earth._is_most_of_it_located_near_the_north_or_south_pole
5. https://www.usgs.gov/special-topic/water-science-school/science/ice-snow-and-glaciers-and-water-cycle?qt-science_center_objects=0#qt-science_center_objects
6. www.livescience.com/29312-great-lakes.htm
7. https://www.netstate.com/states/tables/state_elevation_mean.htm
8. https://www.sfgate.com/columnists/morfordredesign/article/Here-are-the-countries-not-in-the-Paris-Climate-11186190.php

9. https://www.reuters.com/news/picture/who-are-the-worlds-biggest-polluters-idusrtxrksi
10. www.lakepedia.com/lake/volta.htm
11. https://www.britannica.com/place/Lake-Kariba
12. www.worldatlas.com/articles/10-largest-desert
13. www.worldatlas.com/articles/10-largest-desert
14. www.worldatlas.com/articles/10-largest-desert
15. www.worldatlas.com/articles/10-largest-desert
16. www.worldatlas.com/articles/10-largest-desert
17. www.worldatlas.com/articles/10-largest-desert
18. https://www.worldatlas.com/articles/religious-demographics-of-china.html
19. https://www.worldatlas.com/articles/religious-composition-of-australia.html
20. http://www.differencebetween.net/miscellaneous/difference-between-sunni-and-shiite-islam/
21. https://www.worldatlas.com/articles/religious-beliefs-and-freedoms-in-saudi-arabia.html
22. https://www.worldatlas.com/articles/religious-beliefs-and-freedoms-in-iran.html
23. https://www.nationsencyclopedia.com/Africa/Egypt-RELIGIONS.html

24. https://hypertextbook.com/facts/2001/DanielChen.shtml
25. https://www.answers.com/Q/How_many_square_miles_of_forest_world_wide
26. https://slate.com/technology/2014/08/atmospheric-co2-humans-put-40-billion-tons-into-the-air-annually.html
27. https://www.carbonpirates.com/blog/how-much-carbon-do-trees-absorb/
28. https://www.answers.com/Q/What_was_the_total_cost_of_building_Hoover_Dam
29. https://www.voiceofsandiego.org/topics/government/change-rein-costs-wildly-expensive-mid-coast-trolley-project/
30. https://www.sutori.com/story/saudi-arabia-desalination-plant--knkJCRayGSZ7b6dboPHJp8dY
31. https://www.thecalifornian.com/story/news/2019/10/23/clean-water-calam-desal-plant-plan-faces-pushback-over-cost/4057830002/
32. https://permies.com/t/3087/Acres-Trees-planted
33. https://militarybenefits.info/2019-defense-budget/